# ROBOT
## COLORING BOOK

© 2020 Giftsala Publishing
All Rights Reserved.

No part of this publication
may be reproduced,
distributed, or transmitted
in any form or by any means,
including photocopying, recording,
or other electronic or mechanical
methods, without the prior written
permission of the publisher, except
in the case of brief quotations
embodied in critical reviews and
certain other noncommercial uses
permitted by copyright law.
For permission requests,
write to the publisher.

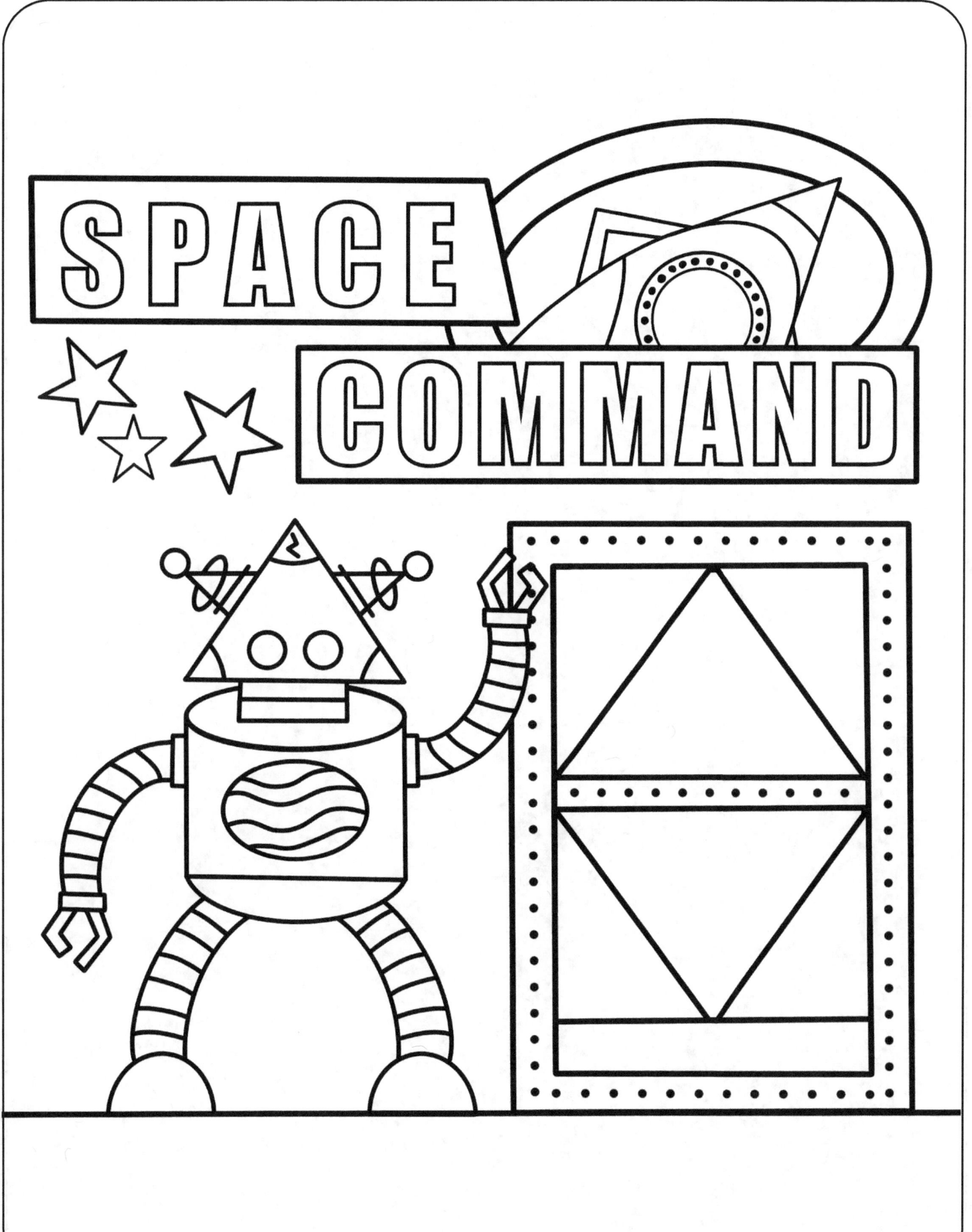

Thanks for purchasing and coloring this activity book! If you enjoyed it, we'd love to hear what you loved most and what you'd like to see in the future. Please consider taking a few minutes to leave a review on Amazon and help others discover us! Being independent artists, it is only with support from amazing people like yourself that we are able to reach more people, share our work and build an amazing community. We can't wait to make more amazing stuff for you.

Thanks for all your support,
Giftsala Publishing

# WANT MORE FUN STUFF?

Join our email list here for a free printable coloring book, tons of freebies, giveaways, flash sale alerts and more fun!
giftsala247@gmail.com

Want some printable fun or want to request something custom made just for you? Visit us on Web
www.giftsala.com

Join the community and share your art on Facebook.
Facebook.com/Giftsala2020